YOUR KNOWLEDGE HAS VALUE

Wolfgang Illig

Statistical analysis in practice and Evaluation of research results

GRIN Verlag

Bibliografische Information der Deutschen Nationalbibliothek:

Die Deutsche Bibliothek verzeichnet diese Publikation in der Deutschen National-
bibliografie; detaillierte bibliografische Daten sind im Internet über http://dnb.d-
nb.de/ abrufbar.

Imprint:

Copyright © 2011 GRIN Verlag GmbH
Druck und Bindung: Books on Demand GmbH, Norderstedt Germany
ISBN: 978-3-656-03815-3

This book at GRIN:

http://www.grin.com/en/e-book/180784/statistical-analysis-in-practice-and-evalua-
tion-of-research-results

GRIN - Your knowledge has value

Der GRIN Verlag publiziert seit 1998 wissenschaftliche Arbeiten von Studenten, Hochschullehrern und anderen Akademikern als eBook und gedrucktes Buch. Die Verlagswebsite www.grin.com ist die ideale Plattform zur Veröffentlichung von Hausarbeiten, Abschlussarbeiten, wissenschaftlichen Aufsätzen, Dissertationen und Fachbüchern.

Visit us on the internet:

http://www.grin.com/

http://www.facebook.com/grincom

http://www.twitter.com/grin_com

Exercise on

Statistical analysis in practice
and
Evaluation of research results

Prepared by

Wolfgang Illig

September 11, 2011

Table of contents

List of Illustrations and formulas

1. Scope of work

The following chapters deal with the scope of work and containing the following steps:

- Definition of the scope of work
- Development of a purposeful data base
- Specification of methods applied, in each case using one method with one variable (e.g. ANOVA models) and one method with two variables (e.g. multiple regression analyses).
- Analysis of the data base on the basis of the above methods by means of software SPSS from IBM
- Presentation of results
- Conclusions and interpretation of results

The project shall be compiled in the English language and shall be not exceed 20 pages in length. The procedures applied are described in the following chapter.

2. Procedure

In order to solve the tasks listed in the previous chapter, first of all some general terms are explained and the applied methods dealing with the subject described. A conceptual formulation is then presented and the required data base compiled. When the data base is completed, it is analyzed and evaluated by means of the defined methods. In conclusion, the results are presented and interpreted and deductions specified accordingly.

3. General description

The explanations of the following terms are intended to enable easier access to the subject field. Processing of the specified tasks would otherwise only be possible with difficulty, if there were no understanding of what lies behind these tasks.

3.1. Software SPSS from the Firma Company

The company SPSS was founded in 1968 with the development of the program of the same name at the American Stanford University by Norman H. Nie, C. Hadlai (Tex) Hull and Dale Bent. At this time the name SPSS stood as an abbreviation for "Statistical Package for the Social Sciences".

Today SPSS is used solely for the original product, as over the years the functions of the software have further developed and today cannot really be abbreviated as such. The SPSS Company was taken over by IBM in 2009.

Today IBM issues the software SPSS Statistics, a module-based program package for statistical analysis of data. The basis module enables fundamental data management and extensive statistic and graphic data analyses with the most used statistical methods[1]. Further, there are various additional modules that can be added to the basic module.

For the purposes of this paper, the author used the German SPSS Version 19. One of the possible statistical calculations which can be made using the software is, for example, the analysis of variance which is described in detail in the next chapter.

3.2. Analysis of Variance (ANOVA)

ANOVA stands for "Analysis of Variance". This statistics method is used to determine the differences between various conditions/groups and to compare more than two conditions with one another.[2]ANOVA is then used when there is a dependent variable and a factor with three or more levels or several factors (independent variables).[3]The analysis of variance compares mean values of three or more conditions.[4]This is an extension of the T-Test to more than two groups or more than one independent variable (functions also with only two conditions – the results are then identical with the T-Test results).[5]The purpose is to investigate the dependence of one variable on a second variable.

With the help of ANOVA the variance of the data under examination can be separated according to systematic variance (variance arising from experimental manipulation, "treatment effects") and non-systematic variance (variance arising from individual differences and experimental errors).[6]Since variance is in direct relationship to the total square sum, this allocation of the total square sum, also

---

[1]Cf. http://de.wikipedia.org/wiki/SPSS, Current as of 30.07.2011
[2]Cf.Rasch / Friese / Hofmann / Naumann, 2009, page 50, translated from the German
[3]Cf.Litz, 2000, page 122, translated from the German
[4]Cf.Zoefel / Bühl, 2000, page 171, translated from the German
[5]Cf. Rumsey, 2008, pape 181, translated from the German
[6]Cf.Rasch / Friese / Hofmann / Naumann, 2009, page 18, translated from the German

known as variance analysis or abbreviated to ANOVA.[7]ANOVA only confirms whether there is a significant effect, i.e. that there are significant differences in the mean values, but it is not known exactly how the mean values are different to one another.[8]Preconditions for the calculation of an Analysis of Variance are listed as follows:

- A variable based on interval scale level[9]
- Normal apportionment of criterion variables in main unit[10]
- At least one independent variable that enables a group allocation[11]
- Comparison groups must comprise independent random samples[12]

SPSS software offers the calculation of an ANOVA under linear regression. This however, only indicates whether there is a significant effect or if the mean values are significantly differentiated. It does not however, state exactly the mean values for differentiation.

As well as variance analyses with one variable in the SPSS there are also possible calculations with several variables, such as a multiple regression analysis. These are described in more detail in the following chapter.

3.3. MultipleRegression Analysis with two or more variables

Following the description of ANOVA, multiple regression analysis is now described as an analysis with two or more variables. This form of multivariant analysis is different from a one variant analysis in that two or more factors are used for the explanation of the criteria variables.[13]In respect of the ANOVA described in the previous chapter it is also possible that this can be calculated with several variables.

---

[7]Cf.Hatzinger / Nagel, 2009, page 225, translated from the German
[8]Cf. Hermann / Homburg / Klarmann, 2007, page 592, translated from the German
[9]Cf. Schnell / Hill / Esser, 2008, page 147, translated from the German
[10]Cf. Janssen / Laatz, 2010, page 347, translated from the German
[11]Cf. Hermann / Homburg / Klarmann, 2007, page 117, translated from the German
[12]Cf. Janssen / Laatz, 2010, page 347, translated from the German
[13]Cf.anssen / Laatz, 2010, page 367, translated from the German

The most widespread statistical procedure for testing or determining multivariant connections is multiple regression analysis.[14]The aim of this procedure is to set up a relationship between a dependent and one or more independent variables.[15]The procedure is used in particular to describe connections in quantitative terms or to forecast dependent variables.[16]Many practical applications are given when there is a variable y and a number of variables X1,.....,Xp that can be connected to y.[17]The regression proceedings can thus be used, for example to quantify the strength of the connection. Mathematical methods determine a function such that the residua are minimal.[18]Residua represent the basis for an estimate of variance of the disturbance variables close to expectations.[19]The form of the function depends to a far extent on the method used. Linear regression comprises only linear functions or logistic regression takes only logistic functions into account. In the subsequent chapter the linear regression used is the model for the specification that the dependent variable y is a linear combination of the regression coefficient β but not necessarily of the dependent variable x.[20]In order to specify the model parameters the method used is that of the smallest quadrants.[21]

Following the above general outline descriptions, the following chapters now deal with the specification and include the hypothesis.

4. Definition and formulation of the task / hypothesis

In this chapter the specific tasks to be undertaken and the hypotheses are presented and defined. This is necessary in order that later, when data is evaluated, a statement (hypothesis) can be made (hypothesis has been confirmed or refuted). The author is at present occupied with his car. He must decide whether to continue using the present car or whether to exchange it in part payment for a new car. In this respect the author is considering some questions, subsequently defined here as the hypotheses.

---

[14]Cf. Wolf / Best, 2010, page 21, translated from the German
[15]Cf.Gramlich, 2002, page 59, translated from the German
[16]Cf.Raithel, 2008, page 156, translated from the German
[17]Cf. Urban / Mayerl, 2011, page 29. Translated from the German
[18]Cf. Eckstein, 2006, page 91, Translated from the German
[19]Cf. Hackl, 2008, page 68, Translated from the German
[20]Cf. Bornmann, 2003, page 95. Translated from the German
[21]Cf. Bortz / Schuster, 2010, page 199, Translated from the German

4.1. Hypothesis 1 (one variable)

The author of this work assumes that there is a linear connection between the mileage reading in the vehicles and their current selling price. It is assumed that vehicles with a high number of kilometers performed will fall linearly in price. This hypothesis is formulated on the basis of data for 100 vehicles that have been researched in the Internet and the book PASW Statistics from Reinhold Hatzinger and Herbert Nagel. The null hypothesis to hypothesis 1 can thus be defined as follows:

$$\beta = 0$$

Formula1: Null hypothesis to hypothesis 1

According to the definition of hypothesis 1, hypothesis 2 can be set up in the next chapter.

4.2. Hypothesis 2 (two variables)

Further, the author of this work assumes that apart from the mileage of the vehicles, also the number of customer services carried out to the vehicle shows a linear correlation to the sales price of the vehicle. The basis is the same data base as for the examination of hypothesis 1. On the basis of this hypothesis the following null hypothesis can be defined as shown in formula 2:

$$\beta 1 = \beta 2 = 0$$

Formula2: Null hypothesis to hypothesis 2

Having defined hypothesis 2, hypothesis 3 is now described in the next chapter.

4.3. Hypothesis 3 (three variables)

In this hypothesis the author also assumes that as well as the mileage reading of the vehicles and the customer services carried out to the vehicle, the fact of whether the vehicle was kept in a garage or in the open also stands in a linear correlation to the sales price of the vehicle. The same data base is used as for the examination of hypotheses 1 and 2. On the basis of the hypotheses the following null hypothesis, as described in formula 3, can be defined:

$$\beta 1 = \beta 2 = \beta 3 = 0$$

Formula3: Null hypothesis to hypothesis 3

Having established the hypotheses, the next chapter describes the composition of the required data or data base.

5. Composition of the required data base

For the examination of the hypotheses defined in the previous chapter, the data are now compiled and edited and a data base established which can be evaluated. For this the following steps are necessary.

5.1. Specification of the required variables

The required variables for the examination of the hypotheses are to be specified. The values received on the basis of empiric research are then recorded. The author of this paper has decided on the following variables:

- Current price
- Current mileage reading
- Number of customer services carried out
- Garage available
- Color of the vehicle

Data were researched on a total of 100 second-hand lower medium-sized vehicles. All vehicles, including the author's vehicle, were about three years old. Research was carried out my means of the offers of vehicles for sale via diverse Internet portals such as autoscout.de or mobile.de. If data were not available, the information was requested from the sellers by telephone. It will be seen late in the evaluation of the data base whether in fact all the data were necessary. The data collected in this way were now recorded under SPSS.

5.2. Setting up the data structure in SPSS

Since the variables have already been defined, the analogous variables structure was now set up in SPSS. After the entries were completed, this file could be saved under "file", save under" in the save-file format of SPSS. Figure 1 shows the variables structure that was set up:

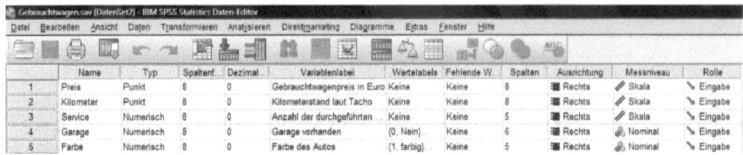

Figure1: variables structure

For the variables price, kilometers and service the measuring level "Scale" was selected and for the two remaining variables, garage and color the measuring Level "Nominal". In addition, for these variables the value label was recorded with possible entry (such as e.g. for garage available) of "yes" or "no". Finally the 100 data sets created were collated one after the other.

With data recorded in this way it was then possible in a further step to carry out the first evaluations or data evaluations.

6. Evaluation of the data base / examination of hypothesis 1

Now that the recorded data base and the variables structure were fixed, the first evaluations could be generated. At the beginning it was examined whether the preconditions as set forth in Chapter 3.2 were given. Then the evaluations for ANOVA could be drawn up.

6.1. Examination of the preconditions

The preconditions set forth in Chapter 3.2 were now examined.

The first precondition, a variable measured at interval scale level was given. An interval scale could be formed through the numeric variable "price", and also the numeric variable "mileage reading".

The next precondition to be examined was the normal spread of the criteria variables. Here there is a normal spread over the variable "mileage reading". The spread of the variable mileage reading represented in figure 2 is given.

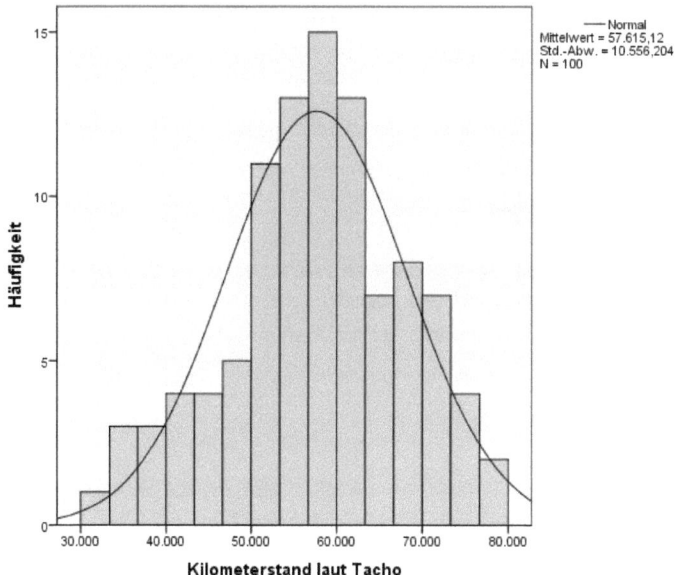

Figure2: Normal spread of the variable mileage reading

The precondition that at least one independent variable which makes group allocation possible is also provided. The variable price can be subdivided into various price classes at any time.

In the view of the author, the independent random samples are also given, as the data of the 100 vehicles were selected randomly from the Internet.

The two box plots that can be seen in the following figure 3, behave in a similar way.

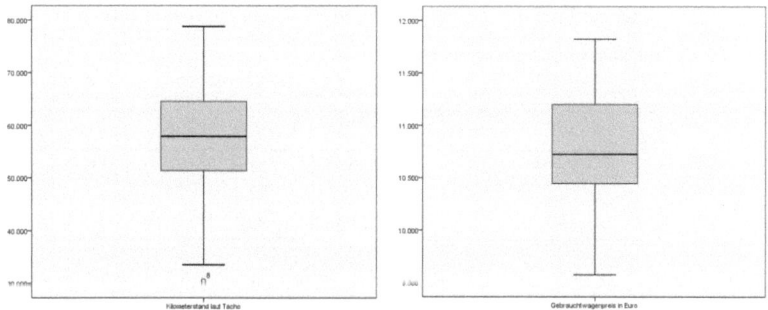

Figure3: Box plots of the variables price and mileage reading

The "box" can be defined as the box in which the mass of the data is to be found.[22] The lines attached to the boxes reflect values which occur less frequently.

Having examined all the preconditions and established that these were met, in the following chapter the linear regression can be calculated with one variable and the hypothesis examined.

6.2. Calculation of regression with one variable and interpretation of the results

After having examined all the preconditions, the calculations on linear regression can be carried out together with ANOVA, the results documented and subsequently interpreted.

For the value Y (response variable) which is to be expected, the following regression equation as shown in formula 4 is set up. Here X stands for the explanatory variable and α or β for the regression coefficients "constant" and "rise" of the regression lines.

$$Y = \alpha + \beta X$$

Formula4: Regression equation for hypothesis 1

For the calculation of regression the function "Analyze" "Regression" and here "Linear" is selected in SPSS. Here the price is selected as the dependent variable and as constant, the predictor variable "kilometers". The significance level 0.05 is then also selected.

After carrying out the calculation, SPSS determines the results shown in figure 4 to the sum of squares of regression.

Modell		Quadratsumme	df	Mittel der Quadrate	F	Sig.
1	Regression	16734110,88	1	16734110,88	182,106	,000ª
	Nicht standardisierte Residuen	9005449,877	98	91892,346		
	Gesamt	25739560,76	99			

ANOVAᵇ

a. Einflußvariablen : (Konstante), Kilometerstand laut Tacho
b. Abhängige Variable: Gebrauchtwagenpreis in Euro

Figure4: ANOVA

---

[22] Cf.Toutenburg / Knöfel, 2007, page 85, translated from the German

The total sum of squares is in the line "total" and is 25,739,560.76; the residua sum of squares in the line "non standardized residua" is 9,005,449,877. The difference between these values is given in the line regression as 16,734,110.88. Based on the comparison of the total sum of squares with the sum of sum of residua squares is a measured value for the quality of the data through a line, the coefficient of determination or R^2.[23] The value for R^2 can be read off in figure 5 Model summary in the column "R-Quadrat". The value is 0.650.

Modellzusammenfassung

Modell	R	R-Quadrat	Korrigiertes R-Quadrat	Standardfehle r des Schätzer s
1	,806[a]	,650	,647	303,138

a. Einflußvariablen : (Konstante), Kilometerstand laut Tacho

Figure5: Model summary

Regression coefficient B and the significance can be taken from the following figure 6 "Coefficients". This figure also contains the test results for the coefficients. The constant is as a rule taken up in the regression equation independently of the significance.[24] The result of the regression coefficients can therefore be read off in the second line.

Koeffizienten[a]

Modell		Nicht standardisierte Koeffizienten		Standardisiert e Koeffiziente n	T	Sig.
		Regressionsk oeffizientB	Standardfehle r	Beta		
1	(Konstante)	13066,766	169,025		77,307	,000
	Kilometerstand laut Tacho	-,039	,003	-,806	-13,495	,000

a. Abhängige Variable: Gebrauchtwagenpreis in Euro

Figure6: Coefficients

The estimate for the regression equation $y = \alpha + \beta X$ is therefore:

$$price = 13.066.766 - 0.039 \times mileage$$

Formula5: Estimate for the regression equation of hypothesis 1

The correlation is negative. The significance of the test (null hypothesis, whether β deviates from "0", although stated as 0.000, only means that the

---

[23]Cf.Hatzinger / Nagel, 2009, page 225, translated from the German
[24]Cf.Hornik / Nagel / Hatzinger, 2011, page 285, translated from the German

significance is very small. The rise is therefore significantly different to "0". Between mileage reading and price of the vehicles there is therefore a significant negative correlation.

After hypothesis 1 was undertaken, the next chapter continues with the examination of hypothesis 2.

7. Evaluation of the data base / examination of hypothesis 2

The collected data base and variables structure can be taken over from the data already used for hypothesis 1. Since here the calculation is made with several variables, the new variables are now examined for their preconditions. The evaluations for the multiple linear regressions are then compiled.

7.1. Examination of preconditions

For the preparation of subsequent calculations the preconditions must be examined once again. Attention is again drawn at this point to the preconditions already examined in chapter 6.1. The focus here should be on the examination of the second explicative variable "service" (on a metric scale).

The scatter diagram price/service as represented in figure 7, in contrast to the scatter diagram price/mileage shows an unusual pattern.

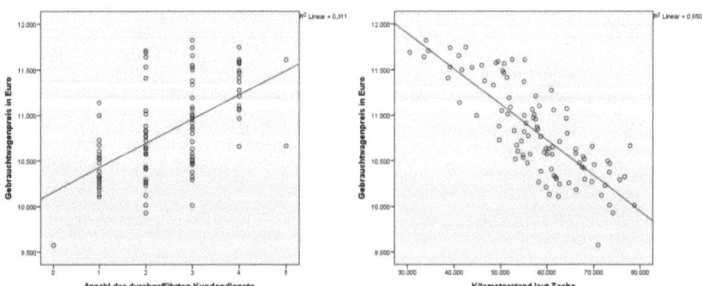

Figure7: Scatter diagram Price / Service and Price / Mileage

The reason for this is presumably that the variable "service" only takes up a few whole number values. The linear adjustment line also incorporated in the illustration, in contrast to the falling adjustment line of the variable "customer service", shows that on average prices rise in accordance with how often the vehicle was taken in for customer service.

After the preconditions have been examined and deemed to be met, then in the following chapter the linear regression for the variables can be calculated and hypothesis 2 examined.

7.2. Calculation of regression with two variables and interpretation of the results

If here too, all the preconditions are given, the calculations for linear regression with two variables can now be undertaken, the results documented and then interpreted.

For the expected value of Y (response variable) the following regression equation represented in formula 6 can be set up. Here X again stands for the explicative variable and α or β again for the regression coefficients "constant" and "rise" of the regression lines.

$$Y = \alpha + \beta 1 X 1 + \beta 2 X 2$$

Formula6: Regression equation for hypothesis 2

For the calculation of regression the function "Analyze" "Regression" and here "Linear" is again selected in SPSS. Here as dependent variable "price" and as constant the "kilometers" as well as the predictor variable "service" are selected. Here once again, the significance level of 0.05 is specified.

After having carried out the calculation, SPSS then determines the result as shown in the following figure 8 for the sum of squares of the linear regression with two variables.

	ANOVA^b					
Modell		Quadratsumme	df	Mittel der Quadrate	F	Sig.
1	Regression	25071994,66	2	12535997,33	1821,530	,000^a
	Nicht standardisierte Residuen	667566,097	97	6882,125		
	Gesamt	25739560,76	99			

a. Einflußvariablen : (Konstante), Anzahl der durchgeführten Kundendienste, Kilometerstand laut Tacho
b. Abhängige Variable: Gebrauchtwagenpreis in Euro

Figure8: ANOVA for two variables

The sum of squares is in the line "total" and is again 25,739,560.76, the residua sum of squares in the line "non standardized residua" is 667,566.097. The difference between these values is given in the line regression as 25,071,994.66. Based on the comparison of the total sum of squares with the sum of sum of residua squares is a measured value for the quality of the data through a line, the coefficient of determination or R^2.[25] The value for R^2 can be read off in figure 9 Model summary in the column "R-Quadrat". The value is 0.974 and comes very close to 1.

Modellzusammenfassung

Modell	R	R-Quadrat	Korrigiertes R-Quadrat	Standardfehle r des Schätzer s
1	,987ª	,974	,974	82,959

a. Einflußvariablen : (Konstante), Anzahl der durchgeführten Kundendienste, Kilometerstand laut Tacho

Figure9: Model summary for two variables

Regression coefficients B and the respective significance can again be taken from the following figure 10 "Coefficients". This figure also contains the test results for the coefficients. As a rule the constants are taken up into the regression equation independently of the significance.[26] The results of the regression coefficients can therefore also be read off from the further lines.

Koeffizientenª

Modell		Nicht standardisierte Koeffizienten		Standardisiert e Koeffiziente n		
		Regressionsk oeffizientB	Standardfehle r	Beta	T	Sig.
1	(Konstante)	12412,257	49,932		248,581	,000
	Kilometerstand laut Tacho	-,039	,001	-,814	-49,788	,000
	Anzahl der durchgeführten Kundendienste	271,675	7,805	,569	34,807	,000

a. Abhängige Variable: Gebrauchtwagenpreis in Euro

Figure10: Coefficients with two variables

Estimate for the regression equation $Y = \alpha + \beta 1 X 1 + \beta 2 X 2$ is therefore:

price = 12.412,257 − 0,039 × mileage + 271,675 × services

Formula7: Estimate for the regression equation to hypothesis 2

The coefficient of determination of R^2 at 0.974 as already mentioned above, is very high and almost reaches the upper limit of 1. From this it can thus be interpreted that more than 97% of the variance can be explained with the two variables

---

[25] Cf.Hatzinger / Nagel, 2009, page 225, translated from the German
[26] Cf.Hornik / Nagel / Hatzinger, 2011, page 285, translated from the German

kilometers and service. The ANOVA (figure 8) gives a highly significant result. In addition the coefficients for kilometers and service show the signs expected according to the scatter diagrams (figure 7). Per driven kilometer the expected price goes down by 3.9 Cent; per service carried out, however, the price increases by approx. 271 Euro.

Now that regression analysis with two variables has been carried out, the next chapter deals with the examination of hypothesis 3 whereby a regression analysis with three variables is calculated.

8. Evaluation of the data base / examination of hypothesis 3

The collected data base and variables structure can be taken over from data already used for hypothesis 1 and 2. Since here too, calculations are made for several variables, the new third variable should be examined and fulfillment of preconditions confirmed. Evaluations for the multiple linear regression are then set up.

8.1. Examination of the preconditions

The preconditions are once again to be examined in preparation for the following calculations. Attention is also drawn at this point to the preconditions already examined in chapters 6.1. and 7.1. The focus now is on examination of the third explicatory variable "garage". Contrary to the previously used metric variables, this is a category variable. This variable is only in the form of "yes" or "no". As coding for these two possibilities, 0 is selected for no and 1 for yes. Through this coding there is a so-called dummy variable for the variable "garage". This procedure with dummy variables is used for the recording of qualitative variables.[27]

In order to examine the correlation of variables "garage" and "price" the box plots price/garage are used, as illustrated in the following figure 7. Here it can quickly be seen that the prices for cars which were kept in a garage are a little higher than for cars without garage facilities.

---

[27]Cf. Von Auer, 2005, page 323, Translated from the German

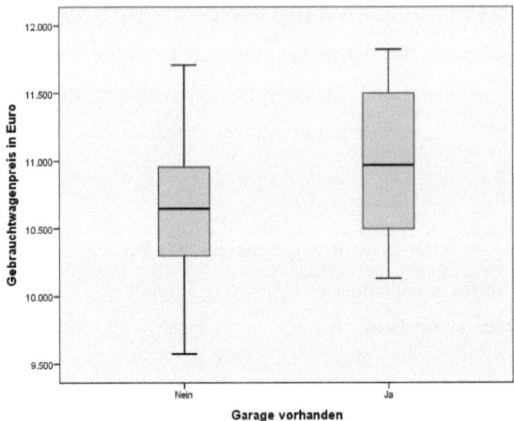

Figure11: Box plots Price / Garage

After the preconditions have been examined and confirmed as met, the linear regression for the three variables can be calculated and hypothesis 3 examined.

8.2. Calculations of regression with three variables and interpretation of the results

After having proven the preconditions, calculations on linear regression with three variables can be made, the results documented and subsequently interpreted.

For the expected value of Y (response variable) the following regression equation can be formed as shown in formula 8. Here X again stands for the explicative variable and α or β again for the regression coefficients "constant" and "rise" of the regression lines.

$$Y = \alpha + \beta 1X1 + \beta 2X2 + \beta 3X3$$

Formula8: Regression equation for hypothesis 3

For the calculation of regression the function "Analyze" "Regression" and here "Linear" is again selected in SPSS. Here as dependent variable "price" and as constant the predictor variable "kilometers", the predictor variables "service" and "garage" are selected. In addition, the significance level is specified once again as 0.05.

After the calculation has been made SPSS determines the result as shown in the following figure 12 for the sum of squares of linear regression with two variables.

ANOVA[b]

Modell		Quadratsumme	df	Mittel der Quadrate	F	Sig.
1	Regression	25105336,94	3	8368445,648	1266,699	,000[a]
	Nicht standardisierte Residuen	634223,815	96	6606,498		
	Gesamt	25739560,76	99			

a. Einflußvariablen : (Konstante), Garage vorhanden, Anzahl der durchgeführten Kundendienste, Kilometerstand laut Tacho
b. Abhängige Variable: Gebrauchtwagenpreis in Euro

Figure12: ANOVA with three variables

The sum of squares is in the line "total" and is again 25,739,560.76, the residua sum of squares in the line "non standardized residua" when compared with calculations with 2 variables, declines and is 634.223.815. Based on the comparison of the total sum of squares with the sum of sum of residua squares a measured value for the quality of the data is through a line, the coefficient of determination or R^2.[28] The value for R^2 can be read off in the model summary as illustrated in figure 13, in the column "R-square". The value is 0.975, comes very close to 1, but in comparison with the calculations with two variables has only changed very slightly.

Modellzusammenfassung

Modell	R	R-Quadrat	Korrigiertes R-Quadrat	Standardfehler des Schätzers
1	,988[a]	,975	,975	81,280

a. Einflußvariablen : (Konstante), Garage vorhanden, Anzahl der durchgeführten Kundendienste, Kilometerstand laut Tacho

Figure13: Model summary for three variables

Regression coefficients B and the respective significance can again be taken from the following figure 14 "Coefficients for 3 variables". This figure also contains the test results for the coefficients. The results of the regression coefficient can therefore also be read off in the further lines.

---

[28]Cf.Hatzinger / Nagel, 2009, page 225, translated from the German

Modell		Nicht standardisierte Koeffizienten		Standardisierte Koeffiziente n		
		Regressionsk oeffizientB	Standardfehle r	Beta	T	Sig.
1	(Konstante)	12374,732	51,695		239,379	,000
	Kilometerstand laut Tacho	-,039	,001	-,806	-48,966	,000
	Anzahl der durchgeführten Kundendienste	269,082	7,734	,564	34,793	,000
	Garage vorhanden	38,015	16,922	,037	2,247	,027

a. Abhängige Variable: Gebrauchtwagenpreis in Euro

Figure14: Coefficients for three variables

Estimate for the regression equation $y = a + \beta_1 X_1 + \beta_2 X_2 + \beta_3 X_3$ is therefore:

price = 12.374,732 − 0,039 × mileage + 269,082 × services + 38,015 × Garage

Formula9: Estimate for the regression equation to hypothesis 3

As already mentioned above, the coefficient of determination of R^2 at 0.975 is very high and almost reaches the upper limit of 1. It can also be interpreted from this that with the three variables of kilometers, service and garage more than 97% of the variance can be accounted for. The ANOVA (figure 12) discloses a highly significant result. Per driven kilometer the expected price decreases by 3.9 Cent and per service carried out the price rises again by approx. 269 Euro. Also the fact that the vehicle was kept in a garage increases the price again by approx. 38 Euro.

9. Conclusion

After the general description and the working out of the statistical principles used, the hypotheses were formulated and then calculated. The user had to spend most of the time for logical understanding of the mathematics involved and for the operation of the SPSS program. Once this was successful, the calculations could be generated relatively easily. All calculations carried out provided the results that the author had more or less presumed roughly beforehand, differing only slightly in the actual final results. Moreover, the intensive preoccupation with linear regression analyses allowed the author to acquire important skills and knowledge that he will be well able to use in his daily work and also in the current program. In addition, using this software also brought the author many benefits compared with previous manual calculations or using Excel.

Bibliography

1.) BORNMANN, LUTZ (2003): Stiftungspropheten in der Wissenschaft. Zuverlässigkeit, Fairness und Erfolg des Peer-Review, 1. Auflage, Münster, WaxmannVerlag, 2003

2.) BORTZ, JÜRGEN / SCHUSTER, CHRISTOF (2010): Statistik für Human- und Sozialwissenschaftler, 7. Auflage, Heidelberg, Springer Verlag, 2010

3.) BROSIUS, FELIX (2006): SPSS für Dummies, 1. Auflage, Weinheim, Wiley-VCH Verlag, 2006

4.) ECKSTEIN, PETER (2006): Repetitorium Statistik: Deskriptive Statistik-Stochastik-Induktive Statistik. MitKlausuraufgaben und Lösungen, 6. Auflage, Wiesbaden, Gabler Verlag, 2006

5.) GRAMLICH, DIETER (2002): Kreditinstitute und Cross Risks, 1. Auflage, Wiesbaden, DeutscherUniversitätsverlag, 2002

6.) HACKL, PETER (2008): Einführung in die Ökonometrie, 1. Auflage, München, Verlag Pearson Studium, 2008

7.) HATZINGER, REINHOLD / NAGEL HERBERT (2009): PASW Statictics, Statistische Methoden und Fallbeispiele, 1. Auflage, München, Verlag Pearson Studium, 2009

8.) HERRMANN, ANDREAS / HOMBURG, CHRISTIAN / KLARMANN, MARTIN (2007): Handbuch Marktforschung: Methoden - Anwendungen – Praxisbeispiele, 3. Auflage, Wiesbaden, Gabler Verlag, 2007

9.) HORNIK, KURT / NAGEL, HERBERT / HATZINGER, REINHOLD (2011): R-Einführung durch angewandte Statistik, 1. Auflage, München, Verlag Pearson Studium, 2011

10.) LITZ, HANS PETER (2000): Multivariate Statistische Methoden: und ihre Anwendung in den Wirtschafts- und Sozialwissenschaften, 1. Auflage, München, OldenbourgWissenschaftsverlag, 2000

11.) RAITHEL, JÜRGEN (2008): Quantitative Forschung. Ein Praxiskurs, 1. Auflage, Wiesbaden, VerlagfürSozialwissenschaften, 2008

12.) RASCH, BJÖRN / FRIESE, MALTE / HOFMANN, WILHELM JOHANN / NAUMANN, EWALD (2009): Quantitative Methoden 2. Einführung in die Statistik für Psychologen und Sozialwissenschaftler, 3. Auflage, Berlin, Heidelberg, New York,Springer Verlag, 2009

13.) RUMSEY, DEBORAH (2008): Weiterführende Statistik für Dummies, 1. Auflage, Weinheim, Wiley-VCH Verlag, 2008

14.) SCHNELL, RAINER / HILL, PAUL B. / ESSER, ELKE (2008): Methoden der empirischen Sozialforschung, 1. Auflage, München, OldenbourgWissenschaftsverlag, 2008

15.) TOUTENBURG, HELGE / KNÖFEL, PHILIPP (2007): Six Sigma: Methoden und Statistik für die Praxis, 1. Auflage, Berlin, Springer Verlag, 2007

16.) URBAN, DIETER / MAYERL, JOCHEN (2011): Regressionsanalyse: Theorie, Technik und Anwendung, 4 Auflage, Wiesbaden, Verlag für Sozialwissenschaften, 2011

17.) VON AUER, LUDWIG (2005): Ökonometrie. Eine Einführung, 3. Auflage, Berlin, Springer Verlag, 2005

18.) ZOEFEL, PETER / BUEHL, ACHIM (2000): Statistik verstehen: Ein Begleitbuch zur computergestützten Anwendung, 2. Auflage, München, Verlag Addison-Wesley, 2000

19.) WIKIPEDIA, Homepage, Stand 30.07.2011
http://de.wikipedia.org/wiki/SPSS

20.) WOLF, CHRISTOF / BEST, HENNIG (2010): Handbuch der sozialwissenschaftlichen Datenanalyse, 1. Auflage, Wiesbaden, VS VerlagfürSozialwissenschaften, 2010